C000134249

WORKS NUMBER:

SITE NAME:

FLOOR/AREA:

DESCRIPTION OF WORKS:

MECHANICAL SITE SUPERVISOR

MECHANICAL CONTRACTS MANAGER

SITE FORMAN

DAY WORK SHEET

WORKS NUMBER:

SITE NAME:

FLOOR/AREA:

DESCRIPTION OF WORKS:

MECHANICAL SITE SUPERVISOR

MECHANICAL CONTRACTS MANAGER

SITE FORMAN

DAY WORK SHEET

WORKS NUMBER:

SITE NAME:

FLOOR/AREA:

DESCRIPTION OF WORKS:

MECHANICAL SITE SUPERVISOR

MECHANICAL CONTRACTS MANAGER

SITE FORMAN

WORKS NUMBER:

SITE NAME:

FLOOR/AREA:

DESCRIPTION OF WORKS:

MECHANICAL SITE SUPERVISOR

MECHANICAL CONTRACTS MANAGER

SITE FORMAN

DAY WORK SHEET

WORKS NUMBER:

SITE NAME:

FLOOR/AREA:

DESCRIPTION OF WORKS:

MECHANICAL SITE SUPERVISOR

MECHANICAL CONTRACTS MANAGER

SITE FORMAN

DAY WORK SHEET

WORKS NUMBER:

SITE NAME:

FLOOR/AREA:

DESCRIPTION OF WORKS:

MECHANICAL SITE SUPERVISOR

MECHANICAL CONTRACTS MANAGER

SITE FORMAN

DAY WORK SHEET

WORKS NUMBER:

SITE NAME:

FLOOR/AREA:

DESCRIPTION OF WORKS:

MECHANICAL SITE SUPERVISOR

MECHANICAL CONTRACTS MANAGER

SITE FORMAN

DAY WORK SHEET

WORKS NUMBER:

SITE NAME:

FLOOR/AREA:

DESCRIPTION OF WORKS:

MECHANICAL SITE SUPERVISOR

MECHANICAL CONTRACTS MANAGER

SITE FORMAN

WORKS NUMBER:

SITE NAME:

FLOOR/AREA:

DESCRIPTION OF WORKS:

MECHANICAL SITE SUPERVISOR

MECHANICAL CONTRACTS MANAGER

SITE FORMAN

DAY WORK SHEET

WORKS NUMBER:

SITE NAME:

FLOOR/AREA:

DESCRIPTION OF WORKS:

MECHANICAL SITE SUPERVISOR

MECHANICAL CONTRACTS MANAGER

SITE FORMAN

DAY WORK SHEET

WORKS NUMBER:

SITE NAME:

FLOOR/AREA:

DESCRIPTION OF WORKS:

MECHANICAL SITE SUPERVISOR

MECHANICAL CONTRACTS MANAGER

SITE FORMAN

DAY WORK SHEET

WORKS NUMBER:

SITE NAME:

FLOOR/AREA:

DESCRIPTION OF WORKS:

MECHANICAL SITE SUPERVISOR

MECHANICAL CONTRACTS MANAGER

SITE FORMAN

DAY WORK SHEET

WORKS NUMBER:

SITE NAME:

FLOOR/AREA:

DESCRIPTION OF WORKS:

_____ MECHANICAL SITE SUPERVISOR

_____ MECHANICAL CONTRACTS MANAGER

_____ SITE FORMAN

DAY WORK SHEET

WORKS NUMBER:

SITE NAME:

FLOOR/AREA:

DESCRIPTION OF WORKS:

MECHANICAL SITE SUPERVISOR

MECHANICAL CONTRACTS MANAGER

SITE FORMAN

DAY WORK SHEET

WORKS NUMBER:

SITE NAME:

FLOOR/AREA:

DESCRIPTION OF WORKS:

MECHANICAL SITE SUPERVISOR

MECHANICAL CONTRACTS MANAGER

SITE FORMAN

WORKS NUMBER:

SITE NAME:

FLOOR/AREA:

DESCRIPTION OF WORKS:

MECHANICAL SITE SUPERVISOR

MECHANICAL CONTRACTS MANAGER

SITE FORMAN

DAY WORK SHEET

WORKS NUMBER:

SITE NAME:

FLOOR/AREA:

DESCRIPTION OF WORKS:

MECHANICAL SITE SUPERVISOR

MECHANICAL CONTRACTS MANAGER

SITE FORMAN

DAY WORK SHEET

WORKS NUMBER:

SITE NAME:

FLOOR/AREA:

DESCRIPTION OF WORKS:

MECHANICAL SITE SUPERVISOR

MECHANICAL CONTRACTS MANAGER

SITE FORMAN

DAY WORK SHEET

WORKS NUMBER:

SITE NAME:

FLOOR/AREA:

DESCRIPTION OF WORKS:

MECHANICAL SITE SUPERVISOR

MECHANICAL CONTRACTS MANAGER

SITE FORMAN

DAY WORK SHEET

WORKS NUMBER:

SITE NAME:

FLOOR/AREA:

DESCRIPTION OF WORKS:

MECHANICAL SITE SUPERVISOR

MECHANICAL CONTRACTS MANAGER

SITE FORMAN

DAY WORK SHEET

WORKS NUMBER:

SITE NAME:

FLOOR/AREA:

DESCRIPTION OF WORKS:

MECHANICAL SITE SUPERVISOR

MECHANICAL CONTRACTS MANAGER

SITE FORMAN

DAY WORK SHEET

WORKS NUMBER:

SITE NAME:

FLOOR/AREA:

DESCRIPTION OF WORKS:

MECHANICAL SITE SUPERVISOR

MECHANICAL CONTRACTS MANAGER

SITE FORMAN

DAY WORK SHEET

WORKS NUMBER:

SITE NAME:

FLOOR/AREA:

DESCRIPTION OF WORKS:

MECHANICAL SITE SUPERVISOR

MECHANICAL CONTRACTS MANAGER

SITE FORMAN

DAY WORK SHEET

WORKS NUMBER:

SITE NAME:

FLOOR/AREA:

DESCRIPTION OF WORKS:

MECHANICAL SITE SUPERVISOR

MECHANICAL CONTRACTS MANAGER

SITE FORMAN

DAY WORK SHEET

WORKS NUMBER:

SITE NAME:

FLOOR/AREA:

DESCRIPTION OF WORKS:

MECHANICAL SITE SUPERVISOR

MECHANICAL CONTRACTS MANAGER

SITE FORMAN

DAY WORK SHEET

WORKS NUMBER:

SITE NAME:

FLOOR/AREA:

DESCRIPTION OF WORKS:

MECHANICAL SITE SUPERVISOR

MECHANICAL CONTRACTS MANAGER

SITE FORMAN

DAY WORK SHEET

WORKS NUMBER:

SITE NAME:

FLOOR/AREA:

DESCRIPTION OF WORKS:

MECHANICAL SITE SUPERVISOR

MECHANICAL CONTRACTS MANAGER

SITE FORMAN

WORKS NUMBER:

SITE NAME:

FLOOR/AREA:

DESCRIPTION OF WORKS:

MECHANICAL SITE SUPERVISOR

MECHANICAL CONTRACTS MANAGER

SITE FORMAN

DAY WORK SHEET

WORKS NUMBER:

SITE NAME:

FLOOR/AREA:

DESCRIPTION OF WORKS:

MECHANICAL SITE SUPERVISOR

MECHANICAL CONTRACTS MANAGER

SITE FORMAN

DAY WORK SHEET

WORKS NUMBER:

SITE NAME:

FLOOR/AREA:

DESCRIPTION OF WORKS:

MECHANICAL SITE SUPERVISOR

MECHANICAL CONTRACTS MANAGER

SITE FORMAN

DAY WORK SHEET

WORKS NUMBER:

SITE NAME:

FLOOR/AREA:

DESCRIPTION OF WORKS:

MECHANICAL SITE SUPERVISOR

MECHANICAL CONTRACTS MANAGER

SITE FORMAN

DAY WORK SHEET

WORKS NUMBER:

SITE NAME:

FLOOR/AREA:

DESCRIPTION OF WORKS:

MECHANICAL SITE SUPERVISOR

MECHANICAL CONTRACTS MANAGER

SITE FORMAN

DAY WORK SHEET

WORKS NUMBER:

SITE NAME:

FLOOR/AREA:

DESCRIPTION OF WORKS:

MECHANICAL SITE SUPERVISOR

MECHANICAL CONTRACTS MANAGER

SITE FORMAN

DAY WORK SHEET

WORKS NUMBER:

SITE NAME:

FLOOR/AREA:

DESCRIPTION OF WORKS:

MECHANICAL SITE SUPERVISOR

MECHANICAL CONTRACTS MANAGER

SITE FORMAN

DAY WORK SHEET

WORKS NUMBER:

SITE NAME:

FLOOR/AREA:

DESCRIPTION OF WORKS:

MECHANICAL SITE SUPERVISOR

MECHANICAL CONTRACTS MANAGER

SITE FORMAN

DAY WORK SHEET

WORKS NUMBER:

SITE NAME:

FLOOR/AREA:

DESCRIPTION OF WORKS:

MECHANICAL SITE SUPERVISOR

MECHANICAL CONTRACTS MANAGER

SITE FORMAN

DAY WORK SHEET

WORKS NUMBER:

SITE NAME:

FLOOR/AREA:

DESCRIPTION OF WORKS:

MECHANICAL SITE SUPERVISOR

MECHANICAL CONTRACTS MANAGER

SITE FORMAN

D A Y   W O R K   S H E E T

WORKS NUMBER:

SITE NAME:

FLOOR/AREA:

DESCRIPTION OF WORKS:

MECHANICAL SITE SUPERVISOR

MECHANICAL CONTRACTS MANAGER

SITE FORMAN

DAY WORK SHEET

WORKS NUMBER:

SITE NAME:

FLOOR/AREA:

DESCRIPTION OF WORKS:

MECHANICAL SITE SUPERVISOR

MECHANICAL CONTRACTS MANAGER

SITE FORMAN

WORKS NUMBER:

SITE NAME:

FLOOR/AREA:

DESCRIPTION OF WORKS:

MECHANICAL SITE SUPERVISOR

MECHANICAL CONTRACTS MANAGER

SITE FORMAN

DAY WORK SHEET

WORKS NUMBER:

SITE NAME:

FLOOR/AREA:

DESCRIPTION OF WORKS:

MECHANICAL SITE SUPERVISOR

MECHANICAL CONTRACTS MANAGER

SITE FORMAN

DAY WORK SHEET

WORKS NUMBER:

SITE NAME:

FLOOR/AREA:

DESCRIPTION OF WORKS:

MECHANICAL SITE SUPERVISOR

MECHANICAL CONTRACTS MANAGER

SITE FORMAN

DAY WORK SHEET

WORKS NUMBER:

SITE NAME:

FLOOR/AREA:

DESCRIPTION OF WORKS:

MECHANICAL SITE SUPERVISOR

MECHANICAL CONTRACTS MANAGER

SITE FORMAN

DAY WORK SHEET

WORKS NUMBER:

SITE NAME:

FLOOR/AREA:

DESCRIPTION OF WORKS:

MECHANICAL SITE SUPERVISOR

MECHANICAL CONTRACTS MANAGER

SITE FORMAN

D A Y   W O R K   S H E E T

WORKS NUMBER:

SITE NAME:

FLOOR/AREA:

DESCRIPTION OF WORKS:

MECHANICAL SITE SUPERVISOR

MECHANICAL CONTRACTS MANAGER

SITE FORMAN

WORKS NUMBER:

SITE NAME:

FLOOR/AREA:

DESCRIPTION OF WORKS:

MECHANICAL SITE SUPERVISOR

MECHANICAL CONTRACTS MANAGER

SITE FORMAN

DAY WORK SHEET

WORKS NUMBER:

SITE NAME:

FLOOR/AREA:

DESCRIPTION OF WORKS:

MECHANICAL SITE SUPERVISOR

MECHANICAL CONTRACTS MANAGER

SITE FORMAN

WORKS NUMBER:

SITE NAME:

FLOOR/AREA:

DESCRIPTION OF WORKS:

MECHANICAL SITE SUPERVISOR

MECHANICAL CONTRACTS MANAGER

SITE FORMAN

DAY WORK SHEET

WORKS NUMBER:

SITE NAME:

FLOOR/AREA:

DESCRIPTION OF WORKS:

MECHANICAL SITE SUPERVISOR

MECHANICAL CONTRACTS MANAGER

SITE FORMAN

DAY WORK SHEET

WORKS NUMBER:

SITE NAME:

FLOOR/AREA:

DESCRIPTION OF WORKS:

MECHANICAL SITE SUPERVISOR

MECHANICAL CONTRACTS MANAGER

SITE FORMAN

DAY WORK SHEET

WORKS NUMBER:

SITE NAME:

FLOOR/AREA:

DESCRIPTION OF WORKS:

MECHANICAL SITE SUPERVISOR

MECHANICAL CONTRACTS MANAGER

SITE FORMAN

WORKS NUMBER:

SITE NAME:

FLOOR/AREA:

DESCRIPTION OF WORKS:

MECHANICAL SITE SUPERVISOR

MECHANICAL CONTRACTS MANAGER

SITE FORMAN

DAY WORK SHEET

WORKS NUMBER:

SITE NAME:

FLOOR/AREA:

DESCRIPTION OF WORKS:

MECHANICAL SITE SUPERVISOR

MECHANICAL CONTRACTS MANAGER

SITE FORMAN

DAY WORK SHEET

WORKS NUMBER:

SITE NAME:

FLOOR/AREA:

DESCRIPTION OF WORKS:

MECHANICAL SITE SUPERVISOR

MECHANICAL CONTRACTS MANAGER

SITE FORMAN

DAY WORK SHEET

WORKS NUMBER:

SITE NAME:

FLOOR/AREA:

DESCRIPTION OF WORKS:

MECHANICAL SITE SUPERVISOR

MECHANICAL CONTRACTS MANAGER

SITE FORMAN

DAY WORK SHEET

WORKS NUMBER:

SITE NAME:

FLOOR/AREA:

DESCRIPTION OF WORKS:

MECHANICAL SITE SUPERVISOR

MECHANICAL CONTRACTS MANAGER

SITE FORMAN

DAY WORK SHEET

WORKS NUMBER:

SITE NAME:

FLOOR/AREA:

DESCRIPTION OF WORKS:

MECHANICAL SITE SUPERVISOR

MECHANICAL CONTRACTS MANAGER

SITE FORMAN

DAY WORK SHEET

WORKS NUMBER:

SITE NAME:

FLOOR/AREA:

DESCRIPTION OF WORKS:

MECHANICAL SITE SUPERVISOR

MECHANICAL CONTRACTS MANAGER

SITE FORMAN

DAY WORK SHEET

WORKS NUMBER:

SITE NAME:

FLOOR/AREA:

DESCRIPTION OF WORKS:

MECHANICAL SITE SUPERVISOR

MECHANICAL CONTRACTS MANAGER

SITE FORMAN

DAY WORK SHEET

WORKS NUMBER:

SITE NAME:

FLOOR/AREA:

DESCRIPTION OF WORKS:

MECHANICAL SITE SUPERVISOR

MECHANICAL CONTRACTS MANAGER

SITE FORMAN

DAY WORK SHEET

WORKS NUMBER:

SITE NAME:

FLOOR/AREA:

DESCRIPTION OF WORKS:

MECHANICAL SITE SUPERVISOR

MECHANICAL CONTRACTS MANAGER

SITE FORMAN

DAY WORK SHEET

WORKS NUMBER:

SITE NAME:

FLOOR/AREA:

DESCRIPTION OF WORKS:

MECHANICAL SITE SUPERVISOR

MECHANICAL CONTRACTS MANAGER

SITE FORMAN

DAY WORK SHEET

WORKS NUMBER:

SITE NAME:

FLOOR/AREA:

DESCRIPTION OF WORKS:

MECHANICAL SITE SUPERVISOR

MECHANICAL CONTRACTS MANAGER

SITE FORMAN

WORKS NUMBER:

SITE NAME:

FLOOR/AREA:

DESCRIPTION OF WORKS:

MECHANICAL SITE SUPERVISOR

MECHANICAL CONTRACTS MANAGER

SITE FORMAN

DAY WORK SHEET

WORKS NUMBER:

SITE NAME:

FLOOR/AREA:

DESCRIPTION OF WORKS:

MECHANICAL SITE SUPERVISOR

MECHANICAL CONTRACTS MANAGER

SITE FORMAN

DAY WORK SHEET

WORKS NUMBER:

SITE NAME:

FLOOR/AREA:

DESCRIPTION OF WORKS:

MECHANICAL SITE SUPERVISOR

MECHANICAL CONTRACTS MANAGER

SITE FORMAN

DAY WORK SHEET

WORKS NUMBER:

SITE NAME:

FLOOR/AREA:

DESCRIPTION OF WORKS:

MECHANICAL SITE SUPERVISOR

MECHANICAL CONTRACTS MANAGER

SITE FORMAN

DAY WORK SHEET

WORKS NUMBER:

SITE NAME:

FLOOR/AREA:

DESCRIPTION OF WORKS:

MECHANICAL SITE SUPERVISOR

MECHANICAL CONTRACTS MANAGER

SITE FORMAN

DAY WORK SHEET

WORKS NUMBER:

SITE NAME:

FLOOR/AREA:

DESCRIPTION OF WORKS:

MECHANICAL SITE SUPERVISOR

MECHANICAL CONTRACTS MANAGER

SITE FORMAN

DAY WORK SHEET

WORKS NUMBER:

SITE NAME:

FLOOR/AREA:

DESCRIPTION OF WORKS:

MECHANICAL SITE SUPERVISOR

MECHANICAL CONTRACTS MANAGER

SITE FORMAN

DAY WORK SHEET

WORKS NUMBER:

SITE NAME:

FLOOR/AREA:

DESCRIPTION OF WORKS:

MECHANICAL SITE SUPERVISOR

MECHANICAL CONTRACTS MANAGER

SITE FORMAN

DAY WORK SHEET

WORKS NUMBER:

SITE NAME:

FLOOR/AREA:

DESCRIPTION OF WORKS:

MECHANICAL SITE SUPERVISOR

MECHANICAL CONTRACTS MANAGER

SITE FORMAN

DAY WORK SHEET

WORKS NUMBER:

SITE NAME:

FLOOR/AREA:

DESCRIPTION OF WORKS:

MECHANICAL SITE SUPERVISOR

MECHANICAL CONTRACTS MANAGER

SITE FORMAN

DAY WORK SHEET

WORKS NUMBER:

SITE NAME:

FLOOR/AREA:

DESCRIPTION OF WORKS:

MECHANICAL SITE SUPERVISOR

MECHANICAL CONTRACTS MANAGER

SITE FORMAN

DAY WORK SHEET

WORKS NUMBER:

SITE NAME:

FLOOR/AREA:

DESCRIPTION OF WORKS:

MECHANICAL SITE SUPERVISOR

MECHANICAL CONTRACTS MANAGER

SITE FORMAN

WORKS NUMBER:

SITE NAME:

FLOOR/AREA:

DESCRIPTION OF WORKS:

MECHANICAL SITE SUPERVISOR

MECHANICAL CONTRACTS MANAGER

SITE FORMAN

DAY WORK SHEET

WORKS NUMBER:

SITE NAME:

FLOOR/AREA:

DESCRIPTION OF WORKS:

MECHANICAL SITE SUPERVISOR

MECHANICAL CONTRACTS MANAGER

SITE FORMAN

DAY WORK SHEET

WORKS NUMBER:

SITE NAME:

FLOOR/AREA:

DESCRIPTION OF WORKS:

MECHANICAL SITE SUPERVISOR

MECHANICAL CONTRACTS MANAGER

SITE FORMAN

DAY WORK SHEET

WORKS NUMBER:

SITE NAME:

FLOOR/AREA:

DESCRIPTION OF WORKS:

MECHANICAL SITE SUPERVISOR

MECHANICAL CONTRACTS MANAGER

SITE FORMAN

DAY WORK SHEET

WORKS NUMBER:

SITE NAME:

FLOOR/AREA:

DESCRIPTION OF WORKS:

MECHANICAL SITE SUPERVISOR

MECHANICAL CONTRACTS MANAGER

SITE FORMAN

D A Y   W O R K   S H E E T

WORKS NUMBER:

SITE NAME:

FLOOR/AREA:

DESCRIPTION OF WORKS:

MECHANICAL SITE SUPERVISOR

MECHANICAL CONTRACTS MANAGER

SITE FORMAN

WORKS NUMBER:

SITE NAME:

FLOOR/AREA:

DESCRIPTION OF WORKS:

MECHANICAL SITE SUPERVISOR

MECHANICAL CONTRACTS MANAGER

SITE FORMAN

DAY WORK SHEET

WORKS NUMBER:

SITE NAME:

FLOOR/AREA:

DESCRIPTION OF WORKS:

MECHANICAL SITE SUPERVISOR

MECHANICAL CONTRACTS MANAGER

SITE FORMAN

DAY WORK SHEET

WORKS NUMBER:

SITE NAME:

FLOOR/AREA:

DESCRIPTION OF WORKS:

MECHANICAL SITE SUPERVISOR

MECHANICAL CONTRACTS MANAGER

SITE FORMAN

DAY WORK SHEET

WORKS NUMBER:

SITE NAME:

FLOOR/AREA:

DESCRIPTION OF WORKS:

MECHANICAL SITE SUPERVISOR

MECHANICAL CONTRACTS MANAGER

SITE FORMAN

DAY WORK SHEET

WORKS NUMBER:

SITE NAME:

FLOOR/AREA:

DESCRIPTION OF WORKS:

MECHANICAL SITE SUPERVISOR

MECHANICAL CONTRACTS MANAGER

SITE FORMAN

DAY WORK SHEET

WORKS NUMBER:

SITE NAME:

FLOOR/AREA:

DESCRIPTION OF WORKS:

MECHANICAL SITE SUPERVISOR

MECHANICAL CONTRACTS MANAGER

SITE FORMAN

DAY WORK SHEET

WORKS NUMBER:

SITE NAME:

FLOOR/AREA:

DESCRIPTION OF WORKS:

MECHANICAL SITE SUPERVISOR

MECHANICAL CONTRACTS MANAGER

SITE FORMAN

DAY WORK SHEET

WORKS NUMBER:

SITE NAME:

FLOOR/AREA:

DESCRIPTION OF WORKS:

MECHANICAL SITE SUPERVISOR

MECHANICAL CONTRACTS MANAGER

SITE FORMAN

DAY WORK SHEET

WORKS NUMBER:

SITE NAME:

FLOOR/AREA:

DESCRIPTION OF WORKS:

MECHANICAL SITE SUPERVISOR

MECHANICAL CONTRACTS MANAGER

SITE FORMAN

DAY WORK SHEET

WORKS NUMBER:

SITE NAME:

FLOOR/AREA:

DESCRIPTION OF WORKS:

MECHANICAL SITE SUPERVISOR

MECHANICAL CONTRACTS MANAGER

SITE FORMAN

WORKS NUMBER:

SITE NAME:

FLOOR/AREA:

DESCRIPTION OF WORKS:

MECHANICAL SITE SUPERVISOR

MECHANICAL CONTRACTS MANAGER

SITE FORMAN

DAY WORK SHEET

WORKS NUMBER:

SITE NAME:

FLOOR/AREA:

DESCRIPTION OF WORKS:

MECHANICAL SITE SUPERVISOR

MECHANICAL CONTRACTS MANAGER

SITE FORMAN

DAY WORK SHEET

WORKS NUMBER:

SITE NAME:

FLOOR/AREA:

DESCRIPTION OF WORKS:

MECHANICAL SITE SUPERVISOR

MECHANICAL CONTRACTS MANAGER

SITE FORMAN

DAY WORK SHEET

WORKS NUMBER:

SITE NAME:

FLOOR/AREA:

DESCRIPTION OF WORKS:

MECHANICAL SITE SUPERVISOR

MECHANICAL CONTRACTS MANAGER

SITE FORMAN

DAY WORK SHEET

WORKS NUMBER:

SITE NAME:

FLOOR/AREA:

DESCRIPTION OF WORKS:

MECHANICAL SITE SUPERVISOR

MECHANICAL CONTRACTS MANAGER

SITE FORMAN

DAY WORK SHEET

WORKS NUMBER:

SITE NAME:

FLOOR/AREA:

DESCRIPTION OF WORKS:

MECHANICAL SITE SUPERVISOR

MECHANICAL CONTRACTS MANAGER

SITE FORMAN

DAY WORK SHEET

WORKS NUMBER:

SITE NAME:

FLOOR/AREA:

DESCRIPTION OF WORKS:

MECHANICAL SITE SUPERVISOR

MECHANICAL CONTRACTS MANAGER

SITE FORMAN

DAY WORK SHEET

WORKS NUMBER:

SITE NAME:

FLOOR/AREA:

DESCRIPTION OF WORKS:

MECHANICAL SITE SUPERVISOR

MECHANICAL CONTRACTS MANAGER

SITE FORMAN

Thank you for your purchase.
Suggestions are welcome for
future editions. Please email
info@kaduct.co.uk with your
ideas

Visit the Kaduct store for
customised day works
books and more

Printed in Great Britain
by Amazon